Lost

Lost

John Wilson

orca currents

ORCA BOOK PUBLISHERS

Library and Archives Canada Cataloguing in Publication

Wilson, John (John Alexander), 1951–, author
Lost / John Wilson.

Issued in print and electronic formats.
ISBN 978-1-4598-1195-9 (paperback).—ISBN 978-1-4598-1196-6 (pdf).—
ISBN 978-1-4598-1197-3 (epub)

I. Title.
PS8595.I5834L657 2016 jC813'.54 C2015-904502-9
C2015-904503-7

First published in the United States, 2015
Library of Congress Control Number: 2015946247

Summary: In this high-interest novel for young readers, teen sleuths Sam and Annabel solve a mystery in the Arctic that has ties to the Franklin Expedition.

Orca Book Publishers is dedicated to preserving the environment and has printed this book on Forest Stewardship Council® certified paper.

Orca Book Publishers gratefully acknowledges the support for its publishing programs provided by the following agencies: the Government of Canada through the Canada Book Fund and the Canada Council for the Arts, and the Province of British Columbia through the BC Arts Council and the Book Publishing Tax Credit.

Cover photography by Tom Gross
Author photo by Katherine Gordon

ORCA BOOK PUBLISHERS
www.orcabook.com

Printed and bound in Canada.

19 18 17 16 • 4 3 2 1

For Russell Potter, William Battersby, Kat Stoetzel, Regina Koellner, Tom Gross and all the other Franklin aficionados who would love to find an oilskin-wrapped journal on an island in Queen Maud Gulf.

Chapter One

"It's spam, Sam," Annabel says without looking up from her crossword puzzle. "Real people don't send email like that. Delete it. Now here's an easy clue for you. 'Two girls, one on each knee.' Seven letters."

"I have no idea," I say. Annabel is teaching me to do cryptic crosswords. I can do regular crosswords, but the clues in cryptic crosswords don't make sense.

Annabel gives me a *how can you be so dumb* look. "Patella," she says. I stare back blankly. "Pat and Ella are girl's names, and a kneecap is called a patella. Simple."

"For you, maybe, but you're weird," I say. Annabel smiles as if I've paid her a compliment. "Sometimes I wish you'd go back to learning pi to some crazy number. At least I could understand that. But come see this email. It's not the usual spam. There's no link to click. They're not asking for anything. It says I've been selected for a free cruise."

"Yippee," Annabel says. "A cruise in the Bahamas. That's useful when you live in Australia."

"It says flights are included. And it's not a cruise to the Bahamas. It goes through the Northwest Passage. It's called 'In the Footsteps of Sir John Franklin'."

A moment later, Annabel is reading the email aloud over my shoulder.

Dear Sam:
ENIGMA TOURS, a division of the Crype Foundation, has a long history of guiding small groups of adventurous souls to locations of extraordinary interest.

ENIGMA TOURS is creating several new adventures based on the world's great mysteries. For our first tour, we are planning something really special, and we hope that you will be interested in a chance to participate.

In 1845, Sir John Franklin led the greatest Arctic expedition ever into the fabled Northwest Passage. Not one of the 129 men on Franklin's two ships lived to see home again. And today their bones lie scattered on the icy shores of Canada's Arctic islands. What went wrong? No one knows. Would you like to be the one to solve the mystery?

ENIGMA TOURS will be offering a small group of people free passage on a trial run of the tour. This will include return airfare and two weeks on our luxury motor yacht Arctic Spray. *All you have to do is spread the word about our great product and allow us to use your name and image in our promotion. Your passage will include a companion.*

A brochure has been mailed, and within a few days I will call you with more information and answer any questions you may have. If you are still interested, your name will be entered in a draw.

I do hope you will consider our unique travel adventures. Thank you for your time.

Sincerely,
Moira Rawdon
Vice-President, ENIGMA TOURS

ENIGMA TOURS
Travel on a ship of the desert.

See the first light of dawn.
Visit the sites of ancient conflicts.
Cross unimagined rivers.

"You're right," Annabel says. "That's not normal spam."

"Do you think it's real?" I ask.

"Could be, I suppose. I can't see any way they could scam you, unless…"

"Unless what?"

"Unless," Annabel goes on, "they're aliens. They might have a plan to abduct teens for a colony on one of Jupiter's moons."

It takes me a second to absorb what she's said. "That's crazy," I tell her.

"Hmmm. I guess you're right," Annabel agrees. "It's much more likely they want you for medical experiments."

"Well, I think it's real," I say, annoyed that Annabel isn't taking the email seriously. I turn back to my computer and search for Enigma Tours.

I click on *News* and see an announcement of upcoming tours.

ON THE TRAIL OF COLONEL FAWCETT: Through the Amazon jungle in search of the lost city of Z

THE MUMMIES OF TURIM: Who are these perfectly preserved bodies in remote Asia?

"Lost cities in the jungle and mummies in the desert sound pretty cool."

"That's how scams work," Annabel explains. "They sound cool and offer something you want to believe in. *Then* they bleed you for money."

"Their website looks pretty good," I say, pointing at the screen. We look at beautiful photographs of luxury ships, trains and camper vans, complete with obviously rich people enjoying themselves. We watch a video of smiling holidaymakers sipping champagne on beaches and equally happy groups being led through ancient ruins.

"It looks real," I say.

"It's easy enough to make a website look fancy," Annabel says.

"Why do you have to be so negative?" I ask, my annoyance returning.

Annabel tilts her head and looks at me. "Not negative," she says. "Just careful. How do you think they found you? And I wonder what the Crype Foundation is. Sounds shady."

"It's easy to get someone's email," I say.

"Okay, but they sent you mail. And they are going to phone you. Street address and cell numbers are tougher to get."

"But still possible," I say. Annabel has a point. It *is* a lot of trouble to go to.

Annabel shrugs. "I guess we're not as private as we like to think."

"Maybe I won't hear from them," I say. My annoyance fades as fast as it appeared. "But I kind of hope it is real. It'd be cool to go to the Arctic."

“Especially in winter,” Annabel says with a laugh.

“That was really bad,” I say. “Even for you. But the timing would be great. It would be awesome to see where they found Franklin’s ship.”

“HMS *Erebus*,” Annabel says. “Did you know that Franklin was lieutenant governor of Tasmania before he went on his last expedition? His wife, Jane, was an explorer in her own right. She was the first European woman to travel across Tasmania.”

“Do you know everything?” I ask.

“Of course not,” Annabel says. “There’s always more to know and learn. I’d quite like to learn more about the Arctic.”

“Would you?” I say. “Don’t forget, Enigma Tours contacted me. I can invite anyone. What’s to say I’ll invite you?”

Annabel punches me on the arm.

“That hurt,” I say.

"You deserved it," she says, looking at her crossword. "Besides, who else would go with you? Now, 'Nothing to hold a spike.' Four letters."

"Nothing...to...hold...a...spike. Four...letters," I repeat slowly. "If nothing is nil and it's holding the letter *a*, the answer could be nail."

Annabel looks impressed. "You're getting the hang of this."

"Me and my friend," I say, turning my laptop to face her. The screen shows *CrypticAid, your crossword helper*.

Annabel shakes her head in despair. "You'll never learn that way."

"Maybe not, but I'll have a lot more spare time."

Chapter Two

The thick, glossy brochure arrives two days later. There are five pages on the cruise through the Northwest Passage. After school Annabel comes over, and we go through it.

"It's very fancy," I say. "Enigma Tours must have lots of money."

"They do, but that doesn't mean they are a good company. We've crossed swords with Humphrey Battleford twice. We know he's rich, but he's also a crook."

"But this brochure *is* well done," Annabel goes on. "It's not too flashy, and the history is accurate and up-to-date. It has the most recent information on the *Erebus* discovery."

"Do you think they'll find Franklin's body on board the ship?" I ask.

"Not a chance," Annabel says.

"Why not?" I've spent a lot of time in the past two days reading about Sir John Franklin's lost expedition. It's no use trying to know as much as Annabel, but I have to give it a shot so that I don't feel totally stupid. "We know from the note the survivors left at Victory Point in 1848 that Franklin was already dead. Maybe they tried to bring his body home."

"No," Annabel says, without admitting that I might be right. "When Franklin died, they had no idea how long it would take them to get home. They couldn't leave him lying around."

"But wouldn't the cold in the Arctic have preserved his body?"

"Yes, but the ships had heating systems. Franklin's body would have become unpleasant quickly. Mind you, after the Battle of Trafalgar they took Nelson's body back to Britain in a barrel of brandy. The alcohol preserved the body."

"That's a bit creepy," I say.

"It gets better," Annabel says with a grin. "The story is that the sailors put a tap on the barrel so they could drink the brandy on the way home."

"That's gross. They drank the brandy the dead man was floating in?"

"Sailors were tough in those days. Maybe Parks Canada should search the *Erebus* for a large barrel."

"How do you find out all this stuff?"

"Research," Annabel says. "And then there's the story of the tall white man with the long teeth."

"I *did* research," I say, "but I didn't find anything about a guy with long teeth." Annabel can be exasperating at times. It's like hanging out with a walking Wikipedia.

"It's a story the Inuit told, but no one believed them. One spring a hunting party found a ship in the ice. It was abandoned except for the body of a tall white man with long teeth."

"You're making this up."

"See? You wouldn't have believed the Inuit either," Annabel says. "A lot of Europeans would have looked tall to the Inuit."

"But long teeth?"

"The Arctic is cold and dry. A body would dry out. When that happens, your eyelids pull back and your teeth and gums recede. After a while it looks like you're staring and that you have very long teeth. It all makes sense."

"What happened then?" I ask.

"When the ice melted, the ship sank. The archaeologists found the *Erebus* very close to where the Inuit claim to have seen it."

"Why did no one believe the Inuit stories?"

"A lot of the British explorers were pompous. They believed their way of doing things was the only way. The Inuit didn't have a written language, so the British thought they were making up stories."

"Do you think this cruise will visit the wreck?"

"I doubt it," says Annabel. "There would be nothing to see unless you're a diver. And the researchers aren't saying exactly where it was that they found the wreck."

"To stop people like Humphrey Battleford stealing things from it."

"Exactly. Can you imagine how much something from that wreck would be worth?"

It's weird to think about putting a dollar value on someone's suffering. I try to imagine the horrors Franklin's men must have endured. They were trapped in an alien landscape thousands of miles from home. They were sick and starving, watching their friends die one by one. I shudder at the thought.

My thoughts are interrupted. "Hi, you two," my dad says. "What's up?"

"Hi, Mr. Butler," Annabel says cheerfully. "We've been reading a brochure for a cruise through the Northwest Passage. It seems they would like Sam to go for free."

"Sounds like a scam to me," Dad says. "In my experience, things that seem too good to be true usually are."

“This one’s a bit different,” I say. “This is the email I got the other day.”

I turn my computer so Dad can see the screen. He reads the email and then scans the brochure. “It is different,” he says. “It might be okay. Has anyone called yet?”

I give Annabel a look I hope says, *See? My dad’s not being negative.*

With amazing timing, my cell phone vibrates in my pocket. I don’t recognize the number. “Hello,” I say.

“Could I speak with Sam Butler?” a female voice asks.

“You’re speaking to him.”

“Excellent. Well, congratulations, Sam. My name’s Moira Rawdon. I’m the vice-president of Enigma Tours.”

Chapter Three

"Hello. Um. Nice to talk to you." I stumble through a few words. I'm fighting to stay calm. I really want to go on this cruise. I nod at Annabel and Dad to indicate that this is *the* call.

"I'm so glad I caught you," Moira Rawdon says. "The time difference between Canada and Australia is quite something to work around."

"You're in Canada?"

"Yes, we have an office in Vancouver. Not as fancy as the head office in New York, but it's okay. Has our brochure arrived?"

"Yes, this morning. It's very nice." I almost kick myself for not being able to say anything smart. Annabel moves closer, her head cocked to one side. Dad is watching from the doorway.

"Excellent," Moira goes on. "I hope there was enough information on the Franklin cruise."

"Yes. I did a project on explorers in ninth grade." Shut up! I think. If they give this tour to anyone, it'll be someone mature and intelligent. Not a kid who babbles about school projects. I take a deep breath. "I think Franklin's fate is interesting. And divers found the wreck of the *Erebus* right where the Inuit said it would be." That's better. Annabel rolls her eyes, but I ignore her. "How long will the cruise be?" I ask. I'm trying to take the pressure off me by asking a question.

"Excellent question." *Excellent* seems to be Moira's favorite word. "There are still a few details to be worked out, but it will be about fourteen days. We'll start during the last week of August. Eventually, we plan a full cruise from Seattle or Vancouver all the way 'round to New York. But this is a test run. We'll be doing only the heart of the expedition, from Sachs Harbour to Pond Inlet. Hope you don't mind."

"Not at all," I say. I have no idea where either place is.

"The exact timing will depend on the weather and ice conditions, but every year sailing in the Arctic gets easier. Only about two hundred ships have ever sailed through the Northwest Passage. But this summer, almost forty are registered to make the journey.

And that includes a large luxury cruise liner carrying nine hundred people. Cruising the Northwest Passage is a growth industry, and Enigma Tours aims to get in on the ground floor."

"Excellent," I say before realizing what I have said. I don't ask what would happen if a nine-hundred-passenger ship got stuck in the ice the way Franklin's did. Instead I say, "How many passengers are you planning to take?"

"Six."

"Six! That's not very many."

"Indeed not," Moira says. "Later tours will be larger, but we want to keep the numbers down to begin with."

"I see," I say.

Annabel leans forward and says, "Hello, Ms. Rawdon. My name's Annabel. I'll be Sam's companion if you select him for this cruise. I have a couple of questions." Annabel smiles and threatens me with a punch.

"Lovely to talk to you, Annabel. Go ahead with your questions."

I put my phone on speaker so that we can all hear what is being said.

"Your company is spending a large amount of money taking only six people on this cruise. How can you be sure that the publicity you get will be worth it?"

"Another excellent question. The thing to remember, Annabel, is that Enigma Tours is not taking this trip only to carry six passengers. Before our first fully commercial cruise, we need to take the ship along the route to make sure that everything we have planned is possible. It's inexpensive to add a handful of extra passengers. This run won't include all the luxuries of a full cruise, but it will give a good sense of what the real thing will be like."

"That makes sense," Annabel says. "What would Sam have to do if he wins the tour?"

"Very little. Both of you will need to allow us to use your image in brochures and news articles covering the cruise. We will also ask that Sam be ready for media interviews both before and after the cruise. Nothing too difficult."

"Okay," Annabel says. "One final question. What made you select Sam? And how did you find him? There must be many people you could have chosen closer to home."

"First, let me remind you that no final decision has been made. But Sam fits our needs, and Australia is one of our target markets. Sam's name came up when one of our researchers came across an article in the *Warrnambool Standard*. Apparently, he helped find an artifact that had been stolen from the museum. A large bird, I believe."

"A porcelain peacock," Annabel says, raising her eyebrows at me. "So what's the next step?"

"Within the next two weeks, Sam will be informed of the results of the draw. If he is successful and still willing to go, I shall pass on information regarding flights, insurance and what to take. There will also be a contract to sign. Of course, Sam is welcome to pass the document by a lawyer if there are any doubts."

"This all sounds very organized," Annabel says and steps away.

"Thank you," I say.

"It was my pleasure," Moira says. "I wish you the best of luck in the draw. And if any questions pop into your head after you put the phone down, please don't hesitate to email or call. It was a pleasure talking with you."

"Thank you," I repeat and close my cell. Annabel and I stare at my dad.

"Well," he says, stroking his chin. "She does seem to be efficient and helpful. I'll need to check the contract if you are selected. But I must say, it looks like a wonderful opportunity. I don't suppose you'd take me along."

"I'd love to, Dad, but I think Annabel would hurt me very badly if I didn't take her."

Dad nods. "I thought as much. Good luck." With a smile, he leaves.

"So, do you still think it's a scam?" I ask. "Moira answered all your questions."

"Sort of," Annabel agrees. "Moira sounds nice, but she avoided the question of how they got your contact information. And she didn't really answer my question of why *you* were selected."

"Don't you think I deserve to be selected?" I ask jokingly. Annabel doesn't laugh, so I try again. "She said they saw the article in the Warrnambool newspaper."

"What are the chances of someone in Canada reading the *Warrnambool*

Standard? And even if they did," Annabel goes on before I can respond, "it was a short piece. And the article was more about me because of my connection with the museum." Annabel's dad runs the museum in Warrnambool. "I would have been the logical contact."

"There you go again," I say, my anger and my voice rising. "Pouring cold water on something I like. You don't have to be right all the time. I think this is a real opportunity, and so does my dad. I'm really, really excited about this trip. If you're not, I'll take someone else."

Annabel looks hurt, but she doesn't say anything.

I feel lousy. I've hurt my favorite person in the whole world. And why? I have doubts about this cruise too. I don't want to, but it *does* seem too good to be true. Annabel is only reminding me of my own doubt.

I sigh, step over to Annabel and put my arm around her shoulder. She looks up at me, close to tears. "I'm sorry," I say. "All you're doing is being rational."

"Like Mr. Spock," Annabel says with a weak laugh, wiping her eyes.

I smile and give her a hug. "I need rational sometimes. I'll have to take you with me to keep me grounded."

"That's the most romantic offer of a date I've ever had," Annabel says with a grin. "But I am glad that I won't have to hurt you."

"Me too."

Chapter Four

My cell rings one morning as I'm going into math class. I recognize Moira's number, so I answer.

"Congratulations, Sam," she says. "You and Annabel have been selected to go on our Northwest Passage cruise."

"Awesome!" I say loudly. Several people, including Mr. MacKay, the teacher, look at me. "That's great news," I say quietly. "What's next?"

"When you're ready, Mr. Butler," MacKay says, "we can begin the calculus lesson."

"Sorry," I mumble as I hurry to my seat.

"Sounds like you're busy," Moira says. "I just wanted to give you the good news. I will email with more details. Again, congratulations."

"Thank you," I say. I fight the urge to punch the air in triumph.

Anything MacKay says about calculus slides past me as if my brain is made of Teflon. When the bell goes, I head for the door like an Olympic sprinter.

"Mr. Butler." I skid to a halt in front of MacKay's desk. "You seem very excited. Is your life more interesting than the assignment that is due today?"

"What? Oh, I haven't quite finished it," I say. This isn't entirely true. I've actually forgotten about it. "Can I have an extension, please?"

"I don't normally give extensions." MacKay seems to be talking incredibly slowly. "But since you haven't been late all term, I will give you until the end of the day tomorrow."

"Thank you."

He isn't finished. "I take it that you've had some good news."

"Umm, yes, I've been selected to go on a cruise through the Northwest Passage." I look toward the door and see Annabel standing in the hall.

"The Northwest Passage?"

That's what I said. What else can he want to know?

"I believe, even with global warming, that route is only open in late August and early September."

"Yes," I say, fidgeting. "That's when I'm going."

"Hmm. I know you haven't been here very long, Mr. Butler. But you do know that will be right in the middle of our second winter school term in Australia."

I do know that. It had slipped my mind. "I'll work something out," I say.

"I hope so," MacKay says. "Don't forget that assignment tomorrow."

"I won't. Thank you," I say as I head across the classroom.

"I've been accepted," I say when I'm barely out the door.

"You mean *we've* been accepted," Annabel says. "And hello."

"Hello. I got a phone call just before class."

Annabel doesn't look as thrilled as I am. "You don't still think it's a scam, do you?" I ask.

Annabel shakes her head. "I don't think it's a scam. No one's asked for any money. But I can't shake the thought that it *is* suspicious."

I struggle not to get annoyed. "When the contract comes, Dad will get his lawyer to look it over."

"I'm sure it will be fine," Annabel says. "I looked up the Crype Foundation on Wikipedia."

"And?" I ask nervously.

"Oh, it seems genuine enough. It's an umbrella organization for a number of companies, all of which are legal as far as I can tell."

"So what's the problem?"

"I googled Crype. It's an unusual surname in the Midwest. Or it's a slang term that means marketing and promotion for its own sake—advertising to no purpose. That seems an odd name to call a company."

"So it must be someone's surname or a made-up word. You've been suspicious all the way through, and each time everything's worked like we were told. Not everyone has a mind as complex and devious as yours."

"I suppose," Annabel says. "It's just that if something seems odd to me, I can't help trying to work it out."

"A modern-day Sherlock Holmes," I say. But then something Mr. MacKay said sinks into my brain. "But we do have a real problem. The cruise is right in the middle of term."

"I know," Annabel says, suddenly grinning, "but I'm sure we can work it out. We can do assignments on Franklin, the Arctic and the Inuit. And if there's as much publicity as Moira suggested, I'm sure the school won't mind the attention. In fact, I was speaking to Ms. Fortune this morning. She thinks the assignments are a great idea, and she's prepared to discuss it with the principal."

"But you didn't know I was going to be accepted."

"I never doubted you," Annabel says, linking arms with me.

"So you're looking forward to coming on a cruise with me."

"I'm only preparing for every eventuality."

"You can't control everything," I say.

"But I can control one thing," Annabel says, heading off down the hall. "There's Pavlova cake for dessert in the cafeteria. Last one there buys the other a piece to celebrate."

Chapter Five

It seems as if Annabel and I have spent the past week on planes or in airports. It takes thirty-three hours to go from Adelaide to Vancouver with stops in Brisbane and Los Angeles. Then it's thirteen more hours from Vancouver to Inuvik via Calgary, Edmonton, Yellowknife and Norman Wells. Annabel doesn't seem to mind. But she has a backpack loaded with books about Franklin. She has everything from biographies of the man to horror novels. With her memory, she'll be a world authority on the Franklin mystery by the end of this cruise.

The final leg of our trip north is with Aklak Air. We fly on a Twin Otter from Inuvik to Sachs Harbour via Paulatuk. The plane is full, which means that there are about twenty people on board. Everyone except for us and an older, loudly dressed couple with American accents is a local.

Annabel is in the window seat beside me. "Look," she says cheerfully as she points out the window. "A pingo."

I lean over and peer past her. "It's a hill," I say without enthusiasm.

"It's a special kind of hill. They form in permafrost, where the ground's frozen all year round. It's like a frost heave, but it goes on for decades, until a hill forms."

I barely hear Annabel. Part of me wants to sleep and the other part is taking in the barren landscape. The landscape beneath us seems as much water as solid ground. Lakes of all sizes and shapes are everywhere, like the marks of some horrible disease. Even the rivers seem unable to decide which way they should flow. They meander in twisted patterns.

"We have the largest pingo in the world," someone says. I turn to look at the old guy in the seat across the aisle from me. He's short, powerfully built and wearing unlaced work boots, jeans and a combat jacket. His face is weather-beaten, but his dark eyes sparkle with life. "Name's Jim," he says with a smile. He holds out a work-roughened hand. "What brings you folks up here?"

I feel the bones in my hand grind together as we shake. "We're going on a cruise through the Northwest Passage," I explain. "I'm Sam and this is Annabel."

"Welcome to the north," Jim says. "Been up this way before?"

"First time," I say.

Jim nods as if he'd already worked that out. "I've been up north near sixty years. Been to Edmonton a few times but don't much like it. Awful busy. Inuvik's as big as I need."

"What do you do?" I ask.

"Hunting and trapping and a bit of guiding. As a boy I worked on those radar stations the Yanks built to spot the Russian bombers coming over if a war broke out."

My tired brain can't think of anything to say, but Annabel fills the silence. "You must fly this route a lot."

"Yeah, I know it pretty well. Excitement is in May and November, when we stop in Fort McPherson because the road's closed for break-up and freeze-up."

The plane banks sharply. "That's the city of Sachs Harbour," he says with a laugh, pointing to a collection of buildings scattered along the shore. The landing strip on the hill above the town is as long as the main street. "Population ninety or so. You and your friend will make a difference while you're here. That your ship?" Jim asks as a large, sleek, three-masted yacht comes into view.

"I guess so," I say.

Jim leans over and peers hard out the window. "*Arctic Spray*," he says.

"How do you know that?" I ask. "You can't see the name, let alone read it from here."

"We Inuit are famed for our eyesight. When a polar bear's stalking you, it's well camouflaged. The only thing you can see is its black nose. If you want to escape, you have to be able to see the bear one to two miles away."

"You can see a polar bear's nose two miles away?" I ask in awe.

"Easy," Jim says. "On clear days, two and half or three. Of course, it helps if your rifle has a telescopic sight. And if a luxury yacht anchors nearby while you're sitting at the end of the dock."

I'm confused—until I notice that the other passengers are chuckling quietly. Even Annabel has trouble keeping a straight face. "Very funny," I say as I feel my face flush.

"Don't take it hard," Jim says. "We only tease people we like. Anyway, it looks like you'll be traveling in style. When do you sail?"

"Tomorrow morning," Annabel says.

"You reading about Franklin?" Jim asks, nodding at Annabel's book. It has a gruesome photograph of the dried body of one of the expedition's crew on the cover.

"Yes," Annabel says. "It's about the three bodies that were dug up on Beechey Island."

"The lead-poisoning theory," Jim says. "It's not the only one, you know."

"I know," Annabel says. She's leaning over me, talking intently to Jim. I feel like I'm trapped between two geeks. "I'm trying to find out as much as I can. The cruise we're going on is passing right by where they all died."

"I know a few things that aren't in any of your books," Jim says. "If you're not sailing until tomorrow, drop by my place this evening for some northern hospitality." Jim is interrupted as we bump along the Sachs Harbour runway. "It's the blue house on the left at the south end of town. Can't miss it. I'll feed you some musk ox and tell you stories about some Kabloonas who tried to do what you're doing."

"Kabloonas?" Annabel asks.

"That's what my ancestors called the first white folks who came up this way.

It means 'the people with bushy eyebrows.' Drop by anytime. Door's always open."

"Thank you," Annabel says. "We will."

We file off the plane and collect our bags. We and the older couple are greeted by an enthusiastic woman dressed in expensive pink-and-blue North Face gear. "Hi," she says cheerily. "I'm Moira. Welcome to the start of the Northwest Passage."

Chapter Six

"This is awesome," I say. The main lounge of the *Arctic Spray* is all polished wood and brass. We are sitting around a table laden with a mouth-watering selection of food and drink. "I think this ship's worth more than our house in Adelaide."

"She is a beauty," Moira says. "Now to introductions." She waves an arm at a bearded man in uniform. "This is Captain Phillips. He's in charge. Everyone has to do exactly what he says."

"Or I'll make you walk the plank." There's polite laughter at the weak joke.

"This is Sam and Annabel," Moira goes on. "They've come all the way from Australia. Though Sam is originally from Canada. This is Billy and Martha Edwards, our lucky winners from Texas."

"San Antonio," Billy says with a wave that includes everyone.

"These two gentlemen are Rob Blair and Terry Mortimer. Their situation is a little different. Rob and Terry are members of KARP—Krill Arctic Research Project. They're based at Fort McPherson in the summer. They are along to check out the possibility of doing research on our cruises."

Rob and Terry look uncomfortable, but everyone shakes hands and tucks into the drinks and food. I turn to Rob, who's sitting beside me. "Have you been here a while?" I ask. "You weren't on the plane this afternoon."

Rob looks startled that I'm talking to him, but Terry answers. "We came up on the flight last week so we could do some preparatory research."

"You flew up from Fort McPherson?" Annabel asks.

"Yeah," Terry says. "That's where we're based in the summer."

"Must be interesting work you do," Annabel goes on. "Krill are so vital to the food chain. It's hard to believe that something as small as krill can supply enough food for something as big as a beluga whale."

"Extraordinary," Terry agrees. "Look, love to talk more, but Rob and I have equipment to check. Catch you later."

"Yeah, see you later," Annabel says.

"Nice for you to have a couple of experts to talk with," I say. I don't get a flicker of a smile from Annabel.

Moira joins us. "Good so far?" she asks.

"It's great," I say through a mouthful of smoked salmon. "The *Arctic Spray*'s amazing."

"She's a beautiful ship. I think you'll enjoy the voyage."

"What time do we sail tomorrow?" I ask.

"Sunrise is about 6:00 AM, but it'll be twilight for two hours before that. Captain Phillips says we'll lift anchor about 5:00 AM. If you're early birds, you can watch the sunrise over breakfast as we sail into Amundsen Gulf."

"Sounds cool," Annabel says. "Any chance we can go ashore this evening? We met someone on the plane who invited us to visit."

"No problem," Moira says. "The Zodiac can take you ashore after the pre-sailing briefing. Rob and Terry want to go ashore in any case."

"Great. Thanks. Come on, Sam. Let's go check out our new home."

Grabbing a final handful of appetizers, I follow Annabel on deck. "What do you think of our companions?" she asks when we're alone at the bow.

"They seem okay," I say.

"Rob and Terry?"

"They're a bit awkward, especially Rob. But they're geeky scientists, so..."

"I'm not so sure."

"That they're geeky?"

"No, that they're scientists. There's something odd about them."

"You're not getting paranoid again, are you?" I ask.

"Not paranoid," Annabel says. "Naturally suspicious. Terry said they flew from Fort McPherson on last week's flight."

"Yeah, there's probably only a couple of flights a week up here."

"There are *no* flights from Fort McPherson to Sachs Harbour at this time of year. Jim, on the plane, said they only stopped at Fort McPherson in May and November."

"Maybe there's another airline," I suggest, "or they flew in on a government plane."

"Maybe, but they don't know much about krill either."

"They only agreed with you that it must take a lot of krill to feed a beluga, and it must." I'm not defending Rob and Terry. I think they're strange too. But I'm worried that Annabel's going to question every single minute of the trip.

"It *would* take a lot of krill to feed a beluga," Annabel agrees, "if belugas ate krill. Only whales with filters in their mouths—baleen—eat krill. Belugas are toothed whales. They don't eat krill."

"They're awkward," I say, "maybe they were nervous. They sure left as quickly as they could."

"Yeah," Annabel says, although she doesn't sound convinced.

Chapter Seven

"Jim wasn't kidding when he said that his door was always open," Annabel says. We are on the front porch of the blue house at the end of town. The door is wide open, and the sound of voices comes from inside.

I knock on the doorframe. There's no response.

"Hello," Annabel shouts into the house. Two children about five or six years old burst through the doorway on our right. They skid to a halt and stand staring at us. Jim appears behind them. "Welcome," he says. "You came for musk ox and stories. Come in."

We follow Jim down the hall and into the kitchen. Two men, one young and one old, look up at us from the table and smile. The two kids follow at a safe distance.

"Are you sure we're not intruding?" Annabel asks.

"Intruding," Jim says. "No such thing up here. I even allow my grand-kids in." He makes a face, and the two children run shrieking with pleasure down the hall. "Now, sit down."

Chairs are dragged in and we're introduced. "My son, Joseph." Jim points to the younger man. "He's the one who can't control those kids."

"They take after their grandfather. What can I do?" Joseph says, shaking hands.

"That's Noah, my father." Jim indicates the other man. "He told me all the stories I know. He's so old he lived through most of them." Noah gives us a nod and a toothless grin.

"Our wives are quilting tonight, so it is up to us to entertain." Jim opens the fridge and hauls out a large plate with a roast on it. He carves a pile of slices and places them on the table in front of us. The meat is dark, almost purple, and blood oozes from the center of each slice. "Have some musk ox," Jim orders. "And tell me it's not the best meat you have ever tasted."

I hesitate, but Annabel picks up a slice and takes a bite. As she chews, blood trickles down her chin. "Wow," she says. "That is awesome. It melts in your mouth."

"The trick is to not overcook it," Jim says.

Annabel stuffs in another mouthful. With less enthusiasm, I lift a piece and take a bite. The strong smell is the first thing I notice. It almost stops me, but I keep going. The taste is unusual but not unpleasant. And Annabel's right—it's incredibly tender. I look up to see everyone staring, waiting for my verdict. "Good," I manage, "but I don't think it'll be on the menu at McDonald's anytime soon." Everyone laughs, and I take another mouthful. As if I've given permission, everyone grabs a slice and tucks in.

"So you're going to look for Franklin," Jim says.

"I don't think we'll be looking," I say, "but the cruise promises to visit places where he was. A lot more is being discovered now that the *Erebus* has been found."

Noah snorts loudly and says something under his breath. Jim laughs. “Noah says he wasn’t aware that the Kabloona was ever lost.

“My family originally comes from the east,” Jim explains, “near King William Island, where Franklin’s ships got stuck in the ice. Noah’s great-grandfather, my great-great-grandfather, met some of Franklin’s men.”

“They ate each other,” Noah says.

“What?” I ask, pausing with a slice of musk ox halfway to my mouth.

“These are southerners, Noah,” Jim says. “Don’t be telling your gruesome stories.”

“No, he’s right,” Annabel says. “There are stories of cannibalism.” Noah nods enthusiastically. “Many of the bones from King William Island have marks that could only come from cutting up the bodies.”

I drop the slice of musk ox back on the plate.

"You're spoiling the boy's appetite, Noah," Jim says. Noah gives me his toothless grin.

"The story I want to tell," Jim goes on, "is about the spring when my great-great-grandfather was on a seal-hunting trip." Noah nods again, and Jim turns to Annabel and me. "On the ice near the south shore of King William Island, they met a party of Kabloonas dragging a sled. They talked with them as best they could with signs and a few words. The men said they had come from the west and that their ships had been trapped in the ice. Then the two parties went their own way."

"I've read about that story," Annabel says. "There were about forty men, and they were starving."

Noah is shaking his head. "Only four Kabloonas and not starving," he says. "If starving, they would have eaten the dog."

"Better than each other," I say, and everybody laughs. "Do you know when this happened?"

"The year that the seal hunting was good," Noah answers.

"Back then, my people didn't count the years the same as we do today," Jim explains. "But it's a good question. I have thought much about it. From other stories, and from what the historians have written, my best guess is the spring of 1850."

"That's two years after the note was left at Victory Point," Annabel says.

Jim nods. "Things are never as simple as the books tell us. And Noah has another story."

We all look at the old man, who nods for Jim to continue. "The summer after they met the four men and the dog, Great-Great-Grandfather's family hunted seals among the islands to the west of the Adelaide Peninsula. When the ice broke up, they camped on an island where they found Kabloona graves on a ridge. And there were bones scattered along the sandy shore. They also met a family who said a huge ship had been found in the ice earlier in the year. The family had gone aboard and taken some useful things. But the ship sank when the ice broke up. Only the masts were left above the water."

"Do you know where the island is?" Annabel asks.

"All I know is that it is a small, sandy island with a ridge along its middle. And there were three pillars of stone on the top of the ridge. I've kept an eye out when I've been out that way, but there are a lot of islands thereabouts."

"If the *Erebus* spent the winter nearby," Annabel says, thinking out loud, "those who died on board might be buried on the island. I wonder—"

"If the four men and the dog were the last to leave the ship before it sank," I interrupt, getting drawn into the speculation, "they would have known they weren't coming back to the ship..."

"And they couldn't carry all their records with them," Annabel goes on.

I finish our train of thought. "They might have buried the records on the same island."

"There are many clues," Annabel says. "Cryptic notes, abandoned supplies, weathered bones, stories and now a complete ship. But how does it all fit together? It's like a jigsaw puzzle with so many missing pieces."

"Look," Jim says, waving his finger in the air like he's following a fly.

"I can't see anything," I say.

"It's the Franklin bug," Jim says. "I think you have both been bitten." Everyone laughs. "Now you see what has fascinated thousands of Kabloonas ever since Franklin and his men disappeared."

"It doesn't fascinate you?" I ask.

"It's a good story, but I am not a Kabloona. I don't need to know everything. What we know is what we know, and that is enough. Besides, isn't it good to have some mystery in life?"

Jim tells us more stories about the past. He also tells us tales of his trapline and his hunting. We hear stories about being stranded on ice floes and being stalked by a polar bear. We're about to leave when Jim's wife, mother and daughter-in-law return from quilting. They insist that we have tea and more food. When we eventually escape, with much exchanging of email addresses, I tell him he'll be the first to know when we find the sandy island with three pillars.

"What did you make of his stories?" I ask as we stroll along Sachs Harbour's only street in near darkness.

"Fascinating. I wonder if we're hearing stories that have never been written down."

"That would be so cool," I say. "Do you think there might be important records on that island?"

"Why not? Sand's good to bury stuff in. But Jim said there are dozens of small sandy islands. How would we know—"

"You folks have a good evening?"

I almost collapse with fright when I hear Terry's voice right beside me. "I didn't hear you come up," I say, turning to see him and Rob emerging from the deeper shadows.

"That's us," Terry says. "Experts on krill and masters of sneaking up on people."

Rob snorts out an unpleasant laugh.

I feel strangely uncomfortable, like Terry might have been eavesdropping on our conversation. We walk in silence back to the Zodiac.

Chapter Eight

For almost a week, we cruise along the north coast of Canada. We pass through Amundsen Gulf, along Dolphin and Union Strait, across Coronation Gulf and up Dease Strait. We set off early each morning and anchor in the late afternoon. I spend most of the cruising time reading, keeping a journal and taking photographs for our school project. Annabel mostly reads and takes notes on her library of Franklin books. She tells me the interesting bits to put in my journal, so I'll look smart without having to work for it.

After we anchor each day, we explore using thc Zodiac or kayaks. We've seen seals, walruses, countless birds and the occasional caribou. We have passed and been passed by several other boats, but the landscape is so vast and empty here that after only a few minutes' walk inland, you feel like the only people in the world.

"Here we are, at the heart of the Franklin mystery," I say. Annabel and I are on the deck of the *Arctic Spray*, watching the sunrise and enjoying a breakfast of eggs Benedict, pancakes and orange juice. The mornings are getting colder, and today there's a skim of frost on the ship's rail. We're wearing our warm jackets to keep out the chill wind. Yesterday we sailed across Queen

Maud Gulf and are now anchored off a cluster of low islands. "How are you enjoying the cruise so far?"

Annabel finishes her last mouthful and says, "Life on board ship is okay. The food is amazing, and everything is comfortable. But the company could be better."

"I know what you mean," I say. We've hardly seen Rob and Terry since we left Sachs Harbour. They only appear at meals, and then they scuttle off before anyone can talk to them. "I suppose they're working on their research."

"If their research can be done in their cabins, they could have stayed home. I expected that they'd be off taking readings and collecting samples at every opportunity."

"Maybe the area they're going to study is later in the cruise," I suggest.

"Maybe," Annabel says. "Billy and Martha are nice enough."

"Yeah," I agree, "but they seem to think that the best part of a cruise is sitting on deck dozing."

"And their only topic of conversation is how much better all the other cruises they've been on were."

"That's not fair," I say. "Martha can also complain that there are no onshore tours."

"And Billy keeps telling Moira that it would be a good idea to bring weapons along to kill any game we see."

"I'm sure Moira thinks that's an *excellent* idea," I say, and we both laugh. "If she asks me one more time how I'm doing, or if I need anything, I'll scream."

"Her relentless cheerfulness *is* a bit hard to take. I don't think we're cruise people," says Annabel.

We sit in silence, watching the rising sun paint a bank of clouds to the north vivid shades of pink and orange.

The islands are dark silhouettes against the bright sky. "Do you think the *Erebus* is nearby?" I ask.

"It must be," Annabel says. "It's in Queen Maud Gulf, and Jim said it was a few miles off some low islands."

"Do you think the answer to the mystery is on one of those islands?"

"Could be, but which one? They all look the same from here."

"I know that people have been searching for the answer to the Franklin mystery for 170 years," I say, "but I can't help hoping that we'll find something."

"It might not be as unlikely as you think," Annabel says. "For those 170 years, most people believed that the ships were crushed in 1848 and that everyone died on a hopeless trek down the coast of King William Island. The discovery of the *Erebus* and Jim's stories suggest that at least one ship wasn't crushed and that not everyone died in 1848."

"So?" I ask when Annabel doesn't say any more.

"People search where they expect to find something," she says.

"So they wouldn't have searched near where the *Erebus* was discovered."

"Not as thoroughly as along the shores of King William Island," Annabel points out. "There are dozens of island over there. Any one of them could be the one that Jim's ancestors were on."

"It'd be so cool to find something. I dream of finding an old musket, an officer's sword or skulls."

"I dream of finding a piece of paper. A last message telling us what happened. Trouble is, finding anything would be horribly frustrating."

"How so?"

"Because we couldn't touch it. We'd have to leave everything exactly as we found it for the archaeologists to work on."

"I suppose so," I say, "but it would still be cool."

"Good morning, you two." Moira's voice intrudes from behind us. "How are you doing? Anything you need?"

I choke back a laugh.

"Good morning, Moira," Annabel says. "I don't think we need anything at the moment. We've just had a lovely breakfast."

"Excellent," Moira says, and I cough loudly. "We'll be anchored here for the day. What are your plans?"

"I thought we were heading up to King William Island today," Annabel says.

"That *was* the plan," Moira says, her smile firmly fixed in place. "However, there are reports of ice drifting down the coast of King William Island, and the captain wants to avoid it. You may have noticed that the wind has changed."

"Yes," Annabel says. "It's from the north and a bit colder."

"There's a storm over the Arctic Ocean," Moira explains. "It won't affect us, but it's pushing broken ice in odd directions. Best to be careful."

"Absolutely," Annabel agrees. "We don't want to end up like Franklin's men."

Moira doesn't find this funny. "Enigma Tours takes the safety of our passengers very seriously. We would never put anyone in harm's way."

"Of course not," Annabel backtracks. "I didn't mean to suggest that. Could Sam and I go exploring some of those islands if the *Arctic Spray*'s not going anywhere today?"

"I don't see why not. I'll organize the Zodiac to take you over."

"We could take the kayaks," I suggest. "They're short enough to fit in the Zodiac."

Moira looks uncertain. "I don't think that's such a good idea. I wouldn't want you going off to different islands."

"We'd just use the kayaks to explore the coast of the closest island," Annabel says. "It's easier than walking."

"I suppose that would be all right. I'll tell the chef to make some sandwiches and flasks of tea. Be sure and dress warm."

I resist the temptation to say, "Thanks, Mom," and nod. Moira hurries off.

"Do you know what?" I ask.

"What?"

"It's going to be an excellent day."

Chapter Nine

After we collect our flasks, packed lunch and Annabel's weighty book about the Inuit stories, we head to the stern, where the Zodiac is tied up. To our dismay, Rob and Terry are already there. They are dressed in military camouflage gear and carry backpacks much larger than our daypacks.

"Thought we'd come with you and collect samples," Terry says.

Annabel shrugs like, *What can we do?*

The Zodiac is crowded as we bump our way over to the island. We're hauling the kayaks onshore when a large red helicopter thumps above us. It banks and does a circuit of the island. I see a pale, plump, smiling face wearing round sunglasses in one of the windows. I only get a brief glimpse, but I have a feeling of familiarity. "Do you think that's part of the expedition diving on *Erebus*?" I ask.

"Could be that they check out anyone coming near the dive site," Annabel says, sounding distracted. The machine circles the *Arctic Spray* and then heads off to the north.

The crewman driving the Zodiac puts the motor in reverse. "I'll pick you up in time for dinner," he shouts as he turns and zooms back toward the ship.

Terry lifts his backpack and heads inland. Rob follows him. "What are you collecting today?" Annabel asks.

Rob doesn't stop, but he turns his head. "Plants," he says.

"Talkative, aren't they?" I say. Annabel just grunts.

We spend the morning puttering on the beach and kayaking along the shore. The ground is made up of small, sharp rocks that are uncomfortable to walk on. They're all weathered to the same dull gray color. The main vegetation is round, dark spots of lichen on the rocks and struggling patches of grass here and there. Small areas of startlingly green moss thrive in sheltered hollows and behind bigger rocks.

The wind has picked up by the time we reach the far side of the island. We settle behind a rock and break open the tea and sandwiches. Annabel has been oddly quiet all morning.

“Not a lot of plant life for Rob and Terry to collect,” I say, trying to make conversation. We’ve seen the pair in the distance, wandering along the low, rocky ridges inland or crouching down examining things.

“I’ve been thinking about the Crype Foundation,” Annabel says thoughtfully. “Did you notice that *Crype* is an anagram of *Percy*?”

“Humphrey Battleford’s dog?” My heart sinks as Annabel’s paranoia begins to cloud an otherwise great day. “That’s stretching it. Didn’t you say it also meant something in marketing and is a surname? Isn’t one of those more likely?”

“Maybe,” Annabel says, “but there’s something else. Do you remember the slogans Enigma Tours uses in their brochures?”

“Vaguely,” I say. “They didn’t make much sense. Isn’t one about traveling in a ship in the desert?”

"The first one," Annabel says. "Travel *on* a ship of the desert. Camels are called ships of the desert."

"That makes more sense," I say, wondering where this conversation is going. "One was about seeing the dawn, which we've done most mornings."

Annabel doesn't crack a smile. "The first light of dawn. And there was one about the sites of ancient wars, but I can't remember the last one."

"Crossing unreal rivers," I say.

"Yeah," Annabel says. "Cross *unimagined* rivers. A ship of the desert's a camel. The first light of dawn would be… a beam, a shaft, a ray? Sites of ancient conflicts could be castles, battlefields or wars. Cross a river might be wade or ford. Camel. Beam. Castle. Wade."

"What are you doing?" I ask, wondering if my friend is suddenly losing her marbles. "This isn't one of your cryptic crosswords to solve."

"No," Annabel agrees, "but I think it might be a game. What do you think of when you think of camels?"

"I don't know. Humps," I suggest.

"That's it!" Annabel jumps to her feet, and I become convinced she's crazy. She ticks off her fingers. "Hump. Ray. Battle. Ford."

I stare at Annabel and remember the familiar face in the helicopter. "You're right! It was Battleford in the helicopter! But it can't be. Can it?"

"It explains a lot. He has the money to buy a travel company. The *Arctic Spray* might even be the yacht Battleford had off Warrnambool. We never got a good look at it. It also explains why Rob and Terry lied about their flight to Sachs Harbour and know nothing about krill. They're Battleford's men."

"What about Billy and Martha?"

"I think they're camouflage. It's you and me he wants up here."

"Why?"

"I don't know, but he went to a lot of effort to lure us here." Annabel's brow furrows the way it does when she's thinking hard. She turns around slowly and gasps. I jump up, expecting there to be a polar bear charging toward us, but Annabel's staring out to sea. "There it is," she says, pointing.

I follow her arm and let out a gasp of my own. "The island in Jim's story."

"It has to be. A small island with sandy beaches, a ridge running along the center and three pillars of stone. It's exactly as he described."

"They're not really pillars," I point out. "More like large rocks."

"True," Annabel agrees, "but it might be a poor translation, or maybe they were bigger 170 years ago. What else can it be?"

"You're right. What are we going to do? The Zodiac won't be back until evening, and we'll probably leave early tomorrow."

"We've got the kayaks," Annabel says.

"It's a long way over," I say, staring at the choppy, gray water between us and the island.

"It's not that far," Annabel says, "and the kayaks are stable. It wouldn't take us long to paddle over there."

"What if the wind picks up more?" I look to the north, where dark clouds are building. "We won't be able to get back."

"The Zodiac will come looking for us, and we won't be far away. They'll see us from here. This is our only chance, and you're right about the wind. We should hurry." Annabel strides down to the kayaks. Wondering how my worry about getting stranded has turned into a reason to go over there as quickly as possible, I follow.

At first the paddling is easy, because we're sheltered by the island. I look back to see Terry and Rob running down to the shore, waving their arms frantically. They're shouting something, but I can't make out the words. "Looks like they wanted to come with us," I shout over to Annabel.

"At least they'll know where we are if we get stuck," she shouts back.

Halfway over, the water gets choppier, and the kayaks are tossed about. I struggle to keep my kayak heading into the waves. The water is bitterly cold when it splashes up on my hands and face. Fortunately, my jacket is waterproof.

After about an hour of paddling, we reach the island. My arms ache, and it feels good to get out and stretch.

Once the kayaks are safely above the tide line, I reach down and grab a handful of sand. It's coarse and dry and runs freely through my fingers. "We could be on a beach in Hawaii," I say.

"Not enough palm trees," Annabel responds, looking around at the moss, lichen and patchy grass.

I look back at the island we've left. It looks closer than it seemed when I was paddling. There's no sign of Rob or Terry.

From the corner of my eye I catch a movement, and I look north to the open water past the island. It takes me a moment to realize what I'm seeing. It's a sleek, expensive three-masted motor yacht, and it's cruising away from us. "The *Arctic Spray*'s abandoning us," I manage to choke out.

Chapter Ten

"There must be a reason," Annabel says.

"What?" I ask.

"Battleford's spent a lot of money and effort to get us here. He wouldn't abandon us."

"Okay," I say, "but you still haven't answered my question. Why is the *Arctic Spray* sailing north?"

"I don't know," Annabel admits. "Perhaps another ship is in trouble, and they got a distress call. They have no way of letting us know. Maybe they contacted Rob and Terry, and that's what they were trying to tell us when we set off. Anyway, we can't paddle after the *Arctic Spray*, so we might as well check this island out."

Unable to think of anything better to do, I trudge after Annabel. We head for the highest point in the middle of the island. The ground rises and dips like a series of low waves. The crests of the waves are sandy and support more plants. The troughs of the waves are stony and bare. I plod over the second trough, trying to get my head around the *Arctic Spray* sailing away, and almost step on something whiter than the surrounding rock.

As I try to avoid stepping on what I saw, my foot turns, twisting my ankle.

A needle of pain shoots through my foot. I cry out and sit down abruptly.

"What's the matter?" Annabel asks, hurrying back to my side.

"I twisted my—" I stop. I'm staring at the thing I injured myself trying to avoid. It's about two inches long and fatter at both ends than in the middle. I point speechlessly at it.

Annabel crouches down and peers at the object. "It's a bone," she says softly.

"Yeah," I agree. "It must be from one of Franklin's men."

"Let's not jump to conclusions," Annabel says. "It could be an animal bone—a bear, a musk ox, a walrus. And even if it *is* a human bone, it could be Inuit."

"I suppose so," I say, disappointed by Annabel's rationality. "But it *could be* from one of Franklin's men. I hate to think I twisted my ankle avoiding a walrus bone."

"It could be," she agrees. "How is your ankle?"

With Annabel's help, I stand. "It's sore," I say, taking a couple of hobbling steps, "but I think it'll be okay if we take it slow."

Moving at a slower pace, we can examine the ground more carefully. At first all I see are gray rocks covered in patches of moss and lichen. Gradually, I begin to see more. In one trough, stones are arranged in a circle.

"It's a tent ring," Annabel explains. "The Inuit placed stones around the bases of their summer tents. When they moved on, they took the tents but left the rings."

"How old is it?"

"No way to tell. It could have been here for hundreds of years."

We find a few more scattered bones, which Annabel declares are "probably walrus." She's probably right, but I *really* want the bones to be from Franklin's men.

Close to the summit, I spot a pale gray pebble with a curved edge. I bend for a closer look, and my heart beats faster. "I've found something," I say.

Annabel joins me, and we stare at the broken bowl of a clay pipe. "This must be from one of Franklin's men," I say triumphantly.

"Could be" is all I get from Annabel.

"Unless your walrus smoked a pipe," I say, annoyed again at Annabel's negativity, "it must be."

"Clay pipes were very common 170 years ago. It might be from one of the search expeditions."

"You're wrong," I say, standing up. "You don't know everything." I know I'm being unfair, but my anger's getting the better of me. "All you want to do is show off how smart you are. This is the moment I've dreamed of ever since the first email. And now we're here, on a possibly unexplored island. The answer to the Franklin mystery might be underneath the next rock. I won't let you spoil that."

I stomp my way up the final gradual slope and onto the top of the ridge. The dark clouds overhead throw everything into shadow and paint the world gray. The wind's getting colder, and my ankle hurts. I slump down beside a pile of stones as the first snowflakes drift past.

Annabel putters down below, and I gradually calm down. I know it's dumb to hope that we would step ashore and stumble upon the answer to the Franklin mystery. If it were that obvious, it would have been found by now. I'm partly angry at my own unrealistic hopes.

I haul off my daypack and open the flask of tea. The snowfall is getting heavier, and the warm liquid tastes wonderful. I instantly feel better and rummage in the pack for a sandwich. Annabel joins me as I take the first bite.

"Sorry," I say through a mouthful of ham and cheese. "I'm not angry at you. I'm annoyed at myself. I expected too much. *And* it's starting to snow."

When Annabel doesn't say anything, I look up. Is she angry at my outburst? She's staring, open-mouthed, past my left shoulder. "What?" I ask, twisting around.

Despite my sore ankle, I leap to my feet, heart pounding. In the midst of the pile of stones not two feet from where I'm standing, weathered with age and partly covered with moss, is a grinning human skull.

Chapter Eleven

The snow is falling steadily as we stand and stare at the skull. This *is* one of Franklin's men. There is no doubt. This person suffered horribly to reach a shallow grave on this lonely hilltop. He watched his companions die one by one. As his own terrible end approached, did he think of the home and loved ones he would never see again?

"I wonder who he was," I say quietly.

Annabel shrugs. "At least his comrades cared and were fit enough to give him a burial."

We step forward and crouch beside the skull. There are other bones poking out between the disturbed stones of the grave. The temptation to dig down to see what's underneath is almost overwhelming, but we resist. Annabel points to a small, ragged patch of faded blue material. "A piece of his uniform," she says.

At last we stand up, stretch and drag our eyes away from the grave. The snow is easing, and the sky is brighter to the north. The three pillars we saw are little more than collapsed cairns of larger stones. "Do you think Franklin's men built these cairns?" I ask.

"No idea," Annabel says. "It's possible. It looks like they buried at least one sailor up here. Perhaps the cairns were supposed to mark that."

“Do you think there might be messages under them?”

“Maybe,” Annabel says, “although Arctic explorers often buried important messages beside a cairn, not under it. Cairns were markers used as navigation beacons.” She stops talking and tilts her head to stare at the cairns. “That’s odd,” she says.

“What is?” I can’t see anything unusual.

“When we looked at the cairns from the other island, they looked as if they were in a straight line, but they’re not. The middle one’s farther back, so they form a triangle.”

“What does that mean?”

“I don’t know,” Annabel says. “Maybe nothing. Go and stand halfway between those two cairns,” she instructs.

I do as Annabel tells me to, not minding that she’s being smart. This is suddenly the most exciting day of my life.

Annabel stands halfway between two other cairns. "This would work better if there were three of us," she says, "but we'll try it anyway. Walk slowly toward the cairn across the triangle from you."

Again I do as I'm told. Annabel does the same from where she's standing. We meet in the middle. "Now what do we do?" I ask.

Annabel is already examining the sandy ground at our feet. "Look around for anything unusual."

I kneel and begin scratching in the sand. I find quite a collection of stones before my fingers scrape against something flat. I have a moment of excitement before I push enough sand aside to see that it's only another stone. It's bigger and flatter than the others, about as long as my foot on each side.

I'm about to cover it over when I notice a mark on the surface. I sweep the rest of the sand off and see that it's an arrow. I can't imagine why anyone would scratch an arrow on a rock, so I call Annabel over. When I point out the arrow, she howls and grabs me around the neck so tightly that I think it's a murder attempt.

"Do you know what this is?" she asks excitedly when she lets me go.

"Graffiti?" I guess.

"The British Navy marked all their possessions with a broad arrow."

"Why would they want to own a rock?" I blurt out before I realize the importance of what she's said. "Someone in the British Navy scratched this arrow."

"Yes. Someone scratched this as a sign."

"That they hid something important here?" I shout, getting caught up in Annabel's excitement.

"I think so."

"Can we lift the rock?" I ask, knowing Annabel's rule against touching any archaeological remains.

She stares for an age at the rock and then at me. "We could take a quick look," she says. "If we find nothing, there's no harm done. If we find something, we can replace the rock and tell someone what we've found."

Before Annabel can change her mind, I reach down, shove my fingers under the edge of the rock and lift gently. Nothing happens. I work my fingers as far under the edge as I possibly can and heave. The rock flips over, spraying sand on Annabel, and I fall backward. By the time I sit up, Annabel almost has her head in the hole. She's carefully scooping out sand.

There is a black package in the hole. It's rectangular, about the size and shape of a thick book. "What is it?" I ask.

"I think it's documents," Annabel says in an awestruck whisper.

My heart sinks. "They're useless," I say. "If they were once documents, they're just a solid black lump now. No one will ever be able to read them."

"You're not looking at the actual paper," Annabel says, her hand hovering over the hole. "What you're seeing is oilcloth. It's made from a ship's sail, rubbed with oil to make it waterproof. The documents, if there are any, will be wrapped inside."

"So they might be dry and readable?" My heart rate is speeding up again.

"They might," Annabel says. "They've been wrapped carefully."

For a moment, I think Annabel's going to lift the package out and open it. I wish she would, but she sighs and pulls her hand back. "We could destroy something invaluable."

That's when we hear a distant thumping sound. The snow has completely stopped now, and the clouds have moved on. I scan the horizon until

I see a growing red dot in the sky. "The helicopter's coming back," I say.

"You're certain you saw Battleford in the helicopter?" Annabel asks.

"I wasn't until you worked out the clues," I say. "Now I'm sure."

The helicopter is heading for the island we landed on that morning, where two figures are waving wildly on the beach.

"That must be Rob and Terry," Annabel says. "Battleford's going to pick them up."

"And they'll tell him we're over here," I say. We both look at the package in the hole.

"We can't let Battleford find this," Annabel says. "If he gets his hands on it, it will end up in his collection and no one will ever know the answer to the Franklin mystery."

We watch as the helicopter lands and Rob and Terry climb aboard.

"If the helicopter lands here," Annabel says, her voice suddenly businesslike, "they'll have to use that flat area at the end of the ridge. You go down and keep Battleford away from here as long as possible. I'm going try and make everything look undisturbed."

"Okay," I say and head along the ridge. I wish we hadn't moved the flat stone. It's going to be really hard for Annabel to make it look like no one has disturbed the ground for 170 years. I feel like kicking myself. Our curiosity may have given Humphrey Battleford the most valuable Franklin relic ever.

Chapter Twelve

I duck and shelter my eyes from the minor sandstorm thrown up by the helicopter rotors. When things calm down, I peer between my fingers and see a door open. Humphrey Battleford steps out. He looks out of place in his three-piece suit. Rob and Terry jump out after him. In a crouching run, they move clear of the rotors and stand up.

“We do meet in the strangest places, Sam,” Battleford says, holding out his hand.

I ignore the hand and say, “You didn’t bring Percy with you?”

Battleford smiles and pulls his hand back. “He doesn’t like helicopters. Can’t say I’m very fond of them myself. Terribly noisy, uncomfortable things. Are you enjoying your cruise?”

“It’s very nice,” I say.

“The *Arctic Spray’s* one of my favorite yachts. A shame the cruise has to come to an end.”

“Where’s the *Arctic Spray* gone?” I ask. “I saw her sail north.”

“You probably have many questions, Sam,” Battleford says, “and I will be happy to answer them. But first I think we should go see what’s keeping Annabel. It’s rude not to come and welcome your guests.”

"Why did you bring us on this cruise?" I ask, but Battleford's already walking up the ridge. I watch him go until I feel Terry's hand push firmly on my back.

"We should go up there too," he says.

Annabel has moved over beside the grave. I resist the temptation to glance at where we found the package. "Hello, Annabel," Battleford says, flashing his most charming smile. "I'm so glad that our paths have crossed once more." He glances down at the skull lying beside the grave. "I see you have found something of interest."

To my horror, he lifts the skull from its mossy bed. Annabel gasps and steps toward Battleford, but Rob steps between them.

"You can't do that," I say as Battleford examines the skull. "You're destroying valuable archaeological evidence."

"Not in very good condition," Battleford muses, ignoring me. "Still, it will make an interesting curio."

"That's sick," Annabel says.

"Sick?" Battleford looks from the skull to Annabel. "Because it's a human skull? Every major museum in the world has a huge collection of human remains. The human remains they show the public are often the biggest crowd pullers. Have you never felt the thrill of peering at a skeleton in a Stone Age grave or a mummy in an open coffin in a museum? Is that sick?"

"That's different," Annabel says, without explaining how.

Battleford smiles and turns to Terry. "Let's dig our friend up and see if he was buried with anything interesting."

Annabel and I protest, but all we can do is watch as Rob and Terry pull away the stones to reveal the bones we had seen poking out. As they work,

Battleford talks. "I wish you two could see my collection. I think you would like it very much."

"Your stolen collection, you mean," Annabel says.

"Very few items in my collection are stolen," Battleford says calmly. "At least, not by me. Oh, the source of many of my pieces is highly questionable. But it is not my responsibility to halt the worldwide trade in stolen antiques. In fact, you would be surprised at how many well-known and important people encourage the trade. There are villages in some parts of the world whose economies rely on selling stolen artifacts to tourists. At least I am wealthy enough to preserve my collection in the best-possible conditions."

"Why did you go to all the effort of sending us on this cruise?" I ask. "You couldn't know that we would find this grave."

"Two reasons," Battleford says. "I wanted something from the *Erebus*. But the location of the wreck was secret. So I needed to find where the archaeologists were diving this summer. I also needed to somehow get the divers away from the site.

"My original plan was to take you on the cruise and arrange to have you left on one of these islands. The *Arctic Spray* would then sail to the dive site and say that some passengers had gone missing and request help for a search. As soon as the dive ship came south to join the search, I would be free to dive on the *Erebus*. They may not be the most convincing scientists, but Rob and Terry are highly qualified divers."

"You said that was your original plan," Annabel says. "What went wrong?"

"Poor research on my part, I'm afraid. I assumed that there would be only one ship. In fact, there's a whole flotilla of ships. Not all the boats would leave to help search, so we wouldn't get a chance to dive."

Battleford shrugs. "As you know from our previous meetings, things don't always turn out as one hopes. Fortunately, Terry overheard your conversation about the stories you were told in Sachs Harbour. When he radioed me about this, another option opened up, and here we are."

"But I still don't get why you invited us," I say.

"I'm sure you think me a cold hearted thief," Battleford says with a slight smile, "but that's not the case. I never married, and I have no children. I suppose a psychiatrist would say that my antiquities are my children, but I've never held with all that head-doctor stuff."

For a moment Battleford's smile becomes sad, but then he brightens.

"I have very much enjoyed my previous encounters with you two."

"Enjoyed?" Annabel says. "But we've cost you a lot of money and stopped you from acquiring rare pieces for your collection."

"The money, I don't care about. I have more than I could possibly spend in two lifetimes," Battleford says with a dismissive wave of his hand. "I *do* care about the pieces, and there were times that I was angry with you for interfering. But then I realized that I was getting a lot of pleasure from our encounters."

"Is that why you put the cryptic clues in the Enigma Tours brochure?" Annabel asks.

"You spotted that," Battleford says gleefully. "I'm glad my little game wasn't wasted. You see, normally it is easy to acquire what catches my fancy. I see something, I pay people, and soon it's part of my collection. You are both smart, and I feel years younger when we lock horns and try to outwit each other. I invited you along to see if you could stop me. It seems that you cannot."

"There's nothing here," Terry says. "Just old bones and a few pieces of rag."

"Well," Battleford says, rubbing his chin thoughtfully and looking around, "perhaps we are searching in the wrong place."

I tense up as he strolls over to the space between the three pillars. "There's nothing over there," I blurt out.

"And how would you know that, Sam?" Battleford asks over his shoulder.

"I mean, there are no graves over there," I say hurriedly.

"The best things are not always hidden in graves," Battleford says, kicking idly in the disturbed sand where Annabel tried to cover the hole we found.

“There seems to be an unusual flat rock here. Terry, would you be so kind as to remove it for me?”

My heart sinks as Terry joins Battleford and crouches down. I look over at Annabel. She shrugs and seems less worried than I feel. She’s right—there’s nothing we can do. Either Terry or Rob could easily keep both of us under control while the other dug up the whole island. Battleford was right. He’s beat us this time.

Chapter Thirteen

After Terry moves the rock and scoops the sand out of the hole, he moves aside for Battleford to reach into the hole. There's a triumphant look on Battleford's face when he stands up, holding the oilskin package.

"No, please leave it." I plead. "It's so important. It could be the answer to everything. You must let the scientists look at it."

Battleford's smile is firmly in place as he walks back over to us, clutching the package to his chest. "You are right, of course," he says. "This is important." For a second I think he's going to hand the package to me. Instead, he says, "You overate scientists. They are just as petty as the rest of our species. I promise that if it contains the answer, I will send you an email."

I slump to the ground, close to tears. Battleford has in his greedy hands the most important find in Arctic history. What's worse is that he has it because of Annabel and me. If we hadn't dug around, he wouldn't have known where to look. I feel totally miserable.

Annabel sits beside me and places a comforting arm around my shoulders. "There's nothing we can do," she says.

She's trying to cheer me up, but it's not working.

"Well," Battleford says, "I would love to open this and let you have a look at what you have lost."

I feel Annabel tense beside me—Battleford's gloating has gotten to her. She relaxes as he continues, "But I am sure you understand how fragile the documents might be. You may not agree with what I do, but I am not irresponsible. I'm as keen as anyone to keep this safe. I shall open the package in controlled conditions. I have contacts in the scientific community who, for a generous donation, would be happy to help me preserve this. And I meant what I said. I shall send you an email telling you what I find."

"You won't get away with this," I say bitterly. "We'll tell the police. They'll raid your home."

"I admire the passion of youth," Battleford says, "but I fear you overestimate the police and underestimate my intelligence. I have many homes, some of which I do not advertise openly. And what would you say? That I stole a package from an empty hole in the ground? There is no proof that I was even here. My helicopter will take me to another of my yachts far from here, and the *Arctic Spray* will continue on her perfectly legal cruise. Oh, the police may make a few comforting noises and open a few files, but they will find nothing."

"So you're going to abandon us here?" I ask.

"Abandon? No, nothing so crude. I would hate to think that you two charming kids might end up like our friend over there." Battleford inclines his head toward the scattered bones around the grave. "I do hope you don't get into too much trouble for digging up that grave."

I feel a surge of anger, but I can't think of anything to say before Battleford goes on. "The archaeologists diving on the wreck will get a radio message from a passing ship, saying that two figures were spotted on this island. It shouldn't be too long before you're picked up. But now I really must be going. As always, it has been a pleasure. I look forward to renewing our relationship in the future."

Battleford heads to the helicopter, closely followed by Rob and Terry. We sit in silence as the machine lifts off. It banks and heads west, and I catch a glimpse of a hand waving in one of the windows.

"He won this one," I say glumly. "And—again—there's nothing we can do to get him arrested."

"A lot of money buys you a lot of safety," Annabel says. "But don't be so sure that he won."

It takes a moment for me to realize what she has said. "What do you mean?" I ask, looking up at her.

Very slowly, Annabel removes her daypack. She places it on the ground between us, opens it and gently lifts out a thick book that is almost black with age. Without opening it, Annabel holds it up so I can see the cover. I can just make out ghostly handwriting.

This is the journal of the years 1845 to 1850, written by James Fitzjames, Captain HMS Erebus and last commander of Sir John Franklin's Great Arctic Expedition. Myself, Surgeon Goodsir, Ice Master Reid and the Boy George Chambers, all that are left, will leave tomorrow and head east in search of rescue. Even after such terrible tragedy, I have high hopes of seeing home once more. If I am wrong, please ensure that this document reaches the hands of the British Admiralty.

God have mercy on us all.

I read it three times. This is it. This is it. This is the final message from the last pitiful survivors before they set off in search of a rescue that never came. I reach out and gently touch the corner of the book. In that moment, I almost understand Humphrey Battleford's passion to possess pieces of the past.

"How did you get this?" I ask, unable to tear my eyes from the book.

"I know I couldn't hide signs of digging. Battleford would see where this was hidden. So I unwrapped the oilcloth and put this in my daypack, the last place he would think of looking." Annabel carefully places Fitzjames's journal back in her pack.

"What did you replace it with?"

Annabel's face breaks into a broad grin. "When Battleford opens the oilcloth in his 'controlled conditions,' he will find my book on the Inuit testimony. I hope he learns something from it."

Annabel laughs first, but soon we are both roaring uncontrollably. Tears of joy and relief stream down our cheeks. We've won, and soon, thanks to the words of the long-dead Captain James Fitzjames, the world will finally know the answer to the Franklin mystery.

Chapter Fourteen

"Do you think Battleford will keep his word and let someone know where we are?" I ask. It's been three hours since the helicopter left. There has been no more snow, and the sun is shining, but I don't want to spend the night in this desolate spot.

"I don't see why he wouldn't," Annabel says. "He seems quite fond of us."

"I doubt he will be when he opens the package."

"Yeah," Annabel says with a grin. "I would love to be a fly on the wall when that happens. Do you know what I missed in this encounter with Battleford? Percy."

"Me too. He doesn't know his master's a crook. Although I don't think the archaeologists would like Percy running around with Franklin's thigh bone in his mouth."

"The teeth marks would give them a new theory about what happened—the dog ate them all."

We both laugh. "I still can't believe the answer to the Franklin mystery is in your backpack," I say.

"Pretty cool, huh."

"I'd love to read it."

"If we're abandoned and doomed to starve like Franklin's men, I promise I'll let you read Fitzjames's journal before I cook you for supper."

"Thanks," I say. "That'll make being turned into stew much better. But I don't think we're going to get the chance."

Annabel follows my pointing hand. There's an aluminum boat heading straight for us. "We're saved!" I shout melodramatically.

"Let's go down to the shore and meet them," Annabel says, standing up.

I stay sitting, staring at the boat. "We were never in any danger. We've only had to wait a few hours, but I'm happy to see the boat arrive. Imagine what the last survivors of Franklin's expedition must have felt, waiting for years for help that never came."

Annabel just nods. We walk in silence to meet our rescuers.

"Typical," Jim says as the boat grounds and a ramp drops from the square prow onto the beach. "A couple of hopeless Kabloonas get themselves in trouble and expect the Inuit to come to their rescue."

"Jim! What are you doing here?" I ask.

"Nice welcome," Jim says, strolling down the ramp. "I could always leave if you're not happy to see me."

"We're delighted to see you," Annabel says. "And have we got a story to tell you."

"Better tell it to this guy as well," Jim says, waving an arm at the tall fair-haired man stepping out of the boat's cabin and coming forward. "He's the expert."

"Dave Whyte," the man says, shaking our hands. "I'm with Parks Canada, Underwater Archaeology."

"You're from the dive on the *Erebus*," I say excitedly.

Dave smiles. "Yeah, but don't ask me what we've found this year. I'm not allowed to say anything. We got a message that some passing tourists had spotted two people on an island and thought they might need help."

"We need a ride home," I say.

"I think we can arrange that," Jim says. "Eh, Dave?"

Before Dave can respond, Annabel says, "I'll make a deal with you."

"First time I've heard of people being rescued wanting to make a deal," Jim says, but Dave looks at Annabel with interest.

"If you tell me what you found, I'll show you what we found," Annabel says with a mischievous smile.

"I'm afraid I can't do that," Dave says. "I don't want to lose my job. But I'd like to see what you've found."

Annabel takes Fitzjames's journal out of her pack and carefully passes it over. As Dave reads the cover, his eyes widen. He tries to say something, but his jaw simply hangs open.

Jim, peering over the archaeologist's shoulder, breaks the silence. "Well, Dave, it looks like these two Kabloonas have a story to tell, and I, for one, would love to hear it."

With considerable effort, Dave manages to speak. "You shouldn't have touched this."

"I know," Annabel agrees. "That's part of the story."

Dave takes a deep breath. "Okay. First off, show me where you found this, and then we'll head back to the dive ship. I suspect you'll have a good audience for your story. As for the deal you offered, we've found things on the *Erebus*, but nothing to compare to this. If you're not in a tearing rush to get home—have either of you ever done any diving?"

Chapter Fifteen

"I still can't believe they took us down on a dive to the *Erebus*," Annabel says. It's the end of September. We're home, enjoying the best French fries in our favorite café. Outside, the sun is shining, and there's zero chance of snow.

"It was incredible," I reply. "Of course, the A-plus we got on our project didn't hurt."

“We should thank Battleford.”

“I guess so. Do you think he’ll ever be caught?”

Annabel looks thoughtful as she washes down the last French fry with the dregs of her Coke. “Tough to see how,” she says. “He’s rich and smart, and he always makes sure that nothing can be traced to him.”

Annabel’s right. We told our story and filled out endless police reports, but nothing has been done. Our case isn’t helped by Enigma Tours being a perfectly genuine company that has never employed anyone called Moira Rawdon and is owned by a very conservative group of German businessmen. All mention of the Crype Foundation on Wikipedia has mysteriously disappeared, and the *Arctic Spray*, even though it was noticed by several people at Sachs Harbour, is not listed on any shipping register.

Everyone involved was overwhelmed by our find, regardless of how it happened. The police admitted something strange had happened, but with no leads to follow there was little they could do.

Fitzjames's journal is in good condition and is being analyzed now. Already the media is clogged with stories from it. Apart from what might have happened to Fitzjames, his three companions and their dog, there is no more Franklin mystery.

My phone beeps that I have an incoming message. "It's from Enigma Tours," I say.

Annabel is at my side in an instant, and we read:

Enigma Tours is pleased that you enjoyed your recent cruise with us. We enjoyed having you along. Personally, I am very much enjoying your gift of stories, although it is not quite what I had expected. I sincerely hope that you will consider us for your travel needs in the future.

Sincerely,
Mr. Toby Heatherlup, FD

"Who's Mr. Toby Heatherlup, and what does FD mean?" I ask.

Annabel almost falls off her chair laughing. "You still can't do cryptic crosswords," she says when she calms down. "It's an anagram. Try rearranging the letters of Mr. Toby Heatherlup, FD.'"

"Humphrey Battleford," I say.

"Yes," Annabel says, "and he's still playing games."

Author's Note

Sir John Franklin did lead an expedition into the Northwest Passage in 1845, and it did end in tragedy. Of the 129 men who set off, 105 were still alive in 1848. They did leave a brief note, written by the third in command of the expedition, James Fitzjames, in a tin can under a pile of stones at Victory Point on King

William Island before marching off into history. The traditional story is that they all died on or near the island in 1848.

There are Inuit tales of starving men dragging sleds, of tents filled with bodies, of cannibalism and of a ship trapped in the ice in Queen Maud Gulf. That ship was the *Erebus*, and its wreck, remarkably well preserved, was discovered by Parks Canada underwater archaeologists in September 2014. Study of the wreck will yield information for years to come and fill in gaps in what we know of the expedition. Perhaps one day a search party on a nearby island will stumble upon an oilcloth-wrapped journal or package of documents buried for safekeeping before the last survivors set off to meet their inevitable fate in 1849 or 1850. I hope not. Mysteries are sometimes more fun.

Other Franklin Books by John Wilson

Graves of Ice—The story of cabin boy George Chambers's experiences aboard *Erebus*. (Reading level: grade 4 to 6)

Across Frozen Seas—A boy living in Saskatchewan begins to have strange dreams about the Arctic. (Reading level: grade 4 to 6)

Discovering the Arctic: The Story of John Rae—An illustrated biography of one of the explorers who searched for Franklin. (Reading level: grade 4 to 6)

North with Franklin: The Lost Journals of James Fitzjames—A novel that recreates James Fitzjames's journal. (Reading level: grade 9+)

John Franklin: Traveller on Undiscovered Seas—A biography of Sir John Franklin. (Reading level: grade 9+)

Acknowledgments

Thanks to Melanie for helping Sam and Annabel come to life once more and to all the John Franklin fans for the information they share, especially Tom Gross for the use of his photograph on the cover.

A long-standing fascination with the Franklin Expedition and the 2014 discovery of the *Erebus* gave John Wilson no choice but to write this third mystery featuring Sam and Annabel. John is the author of numerous books for young readers; every one of his books deals with the past. He lives on Vancouver Island, British Columbia. For more information, visit www.johnwilsonauthor.com.

orca *currents*

9781459806986 9.95 PB
9781459807105 16.95 LIB
9781459806993 PDF • 9781459807006 EPUB

Sam and Annabel are visiting Drumheller, Alberta, where they get to hang out on a dinosaur dig. Annabel, an avid learner, wants to spend as much time as she can near the dig, much to Sam's dismay. But when they learn the dig has uncovered scientifically important bones, even Sam is interested. In fact, the whole town is talking. When Sam and Annabel discover that Humphrey Battleford, a famous collector of stolen goods, is in the area, they are on high alert to keep the ancient bones safe.

Titles in the Series

orca currents

121 Express
Monique Polak

Ace's Basement
Ted Staunton

Agent Angus
K.L. Denman

Alibi
Kristin Butcher

Bad Business
Diane Dakers

Bear Market
Michele Martin Bossley

Benched
Cristy Watson

Beyond Repair
Lois Peterson

The Big Apple Effect
Christy Goerzen

The Big Dip
Melanie Jackson

Bio-pirate
Michele Martin Bossley

Blob
Frieda Wishinsky

Bones
John Wilson

Branded
Eric Walters

Cabin Girl
Kristin Butcher

Caching In
Kristin Butcher

Camp Disaster
Frieda Wishinsky

Camp Wild
Pam Withers

Caught in the Act
Deb Loughead

Chat Room
Kristin Butcher

Cheat
Kristin Butcher

Chick: Lister
Alex Van Tol

Cracked
Michele Martin Bossley

Crossbow
Dayle Campbell Gaetz

Daredevil Club
Pam Withers

Destination Human
K.L. Denman

Disconnect
Lois Peterson

Dog Walker
Karen Spafford-Fitz

Explore
Christy Goerzen

Eyesore
Melanie Jackson

FaceSpace
Adrian Chamberlain

Farmed Out
Christy Goerzen

Fast Slide
Melanie Jackson

Finding Elmo
Monique Polak

Flower Power
Ann Walsh

Fraud Squad
Michele Martin Bossley

Hate Mail
Monique Polak

High Wire
Melanie Jackson

Hold the Pickles
Vicki Grant

Horse Power
Ann Walsh

Hypnotized
Don Trembath

In a Flash
Eric Walters

Junkyard Dog
Monique Polak

Laggan Lard Butts
Eric Walters

Leggings Revolt
Monique Polak

Living Rough
Cristy Watson

Lost
John Wilson

Manga Touch
Jacqueline Pearce

Marked
Norah McClintock

Maxed Out
Daphne Greer

Mirror Image
K.L. Denman

Nine Doors
Vicki Grant

On Cue
Cristy Watson

Oracle
Alex Van Tol

Out of Season
Kari Jones

Perfect Revenge
K.L. Denman

Pigboy
Vicki Grant

Power Chord
Ted Staunton

Pyro
Monique Polak

Queen of the Toilet Bowl
Frieda Wishinsky

Rebel's Tag
K.L. Denman

Reckless
Lesley Choyce

Rise of the Zombie Scarecrows
Deb Loughead

See No Evil
Diane Young

Sewer Rats
Sigmund Brouwer

The Shade
K.L. Denman

Siege
Jacqueline Pearce

Skate Freak
Lesley Choyce

Slick
Sara Cassidy

The Snowball Effect
Deb Loughead

Special Edward
Eric Walters

Splat!
Eric Walters

Spoiled Rotten
Dayle Campbell Gaetz

Stolen
John Wilson

Storm Tide
Kari Jones

Struck
Deb Loughead

Stuff We All Get
K.L. Denman

Sudden Impact
Lesley Choyce

Swiped
Michele Martin Bossley

Tampered
Michele Martin Bossley

Three Good Things
Lois Peterson

Vanish
Karen Spafford-Fitz

Watch Me
Norah McClintock

Windfall
Sara Cassidy

Wired
Sigmund Brouwer

orca *currents*

For more information on all the books in the Orca Currents series, please visit **www.orcabook.com.**